POLICE OPERATION

H. Beam Piper

Police Operation

Table of Contents

Police Operation

Police Operation

"...there may be something in the nature of an occult police force, which operates to divert human suspicions, and to supply explanations that are good enough for whatever, somewhat in the nature of minds, human beings have—or that, if there be occult mischief makers and occult ravagers, they may be of a world also of other beings that are acting to check them, and to explain them, not benevolently, but to divert suspicion from themselves, because they, too, may be exploiting life upon this earth, but in ways more subtle, and in orderly, or organised, fashion." Charles Fort: "LO!"

John Strawmyer stood, an irate figure in faded overalls and sweat–whitened black shirt, apart from the others, his back to the weathered farm–buildings and the line of yellowing woods and the cirrus–streaked blue October sky. He thrust out a work–gnarled hand accusingly.

"That there heifer was worth two hund'rd, two hund'rd an' fifty dollars!" he clamored. "An' that there dog was just like one uh the fam'ly; An' now look at'm! I don't like t' use profane language, but you'ns gotta *do* some'n about this!"

Steve Parker, the district game protector, aimed his Leica at the carcass of the dog and snapped the shutter. "We're doing something about it," he said shortly. Then he stepped ten feet to the left and edged around the mangled heifer, choosing an angle for his camera shot.

The two men in the gray whipcords of the State police, seeing that Parker was through with the dog, moved in and squatted to examine it. The one with the triple chevrons on his sleeves took it by both forefeet and flipped it over on its back. It had been a big brute, of nondescript breed, with a rough black–and–brown coat. Something had clawed it deeply about the head, its throat was slashed transversely several times, and it had been disemboweled by a single slash that had opened its belly from breastbone to tail. They looked at it carefully, and then went to stand beside Parker while he photographed the dead heifer. Like the dog, it had been talon–raked on either side of the head, and its throat had been slashed deeply several times. In addition, flesh had been torn from one flank in great strips.

"I can't kill a bear outa season, no!" Strawmyer continued his plaint. "But a bear comes an' kills my stock an' my dog; that there's all right! That's the kinda deal a farmer always gits, in this state! I don't like t' use profane language—"

"Then don't!" Parker barked at him, impatiently. "Don't use any kind of language. Just put in your claim and shut up!" He turned to the men in whipcords and gray Stetsons. "You boys seen everything?" he asked. "Then let's go."

* * * * *

They walked briskly back to the barnyard, Strawmyer following them, still vociferating about the wrongs of the farmer at the hands of a cynical and corrupt State government. They climbed into the State police car, the sergeant and the private in front and Parker into the rear, laying his camera on the seat beside a Winchester carbine.

"Weren't you pretty short with that fellow, back there, Steve?" the sergeant asked as the private started the car.

"Not too short. 'I don't like t' use profane language'," Parker mimicked the bereaved heifer owner, and then he went on to specify: "I'm morally certain that he's shot at least four illegal deer in the last year. When and if I ever get anything on him, he's going to be sorrier for himself then he is now."

"They're the characters that always beef their heads off," the sergeant agreed. "You think that whatever did this was the same as the others?"

"Yes. The dog must have jumped it while it was eating at the heifer. Same superficial scratches about the head, and deep cuts on the throat or belly. The bigger the animal, the farther front the big slashes occur. Evidently something grabs them by the head with front claws, and slashes with hind claws; that's why I think it's a bobcat."

"You know," the private said, "I saw a lot of wounds like that during the war. My outfit landed on Mindanao, where the guerrillas had been active.

And this looks like bolo–work to me."

"The surplus–stores are full of machetes and jungle knives," the sergeant considered. "I think I'll call up Doc Winters, at the County Hospital, and see if all his squirrel–fodder is present and accounted for."

"But most of the livestock was eaten at, like the heifer," Parker objected.

"By definition, nuts have abnormal tastes," the sergeant replied. "Or the eating might have been done later, by foxes."

"I hope so; that'd let me out," Parker said.

"Ha, listen to the man!" the private howled, stopping the car at the end of the lane. "He thinks a nut with a machete and a Tarzan complex is just good clean fun. Which way, now?"

"Well, let's see." The sergeant had unfolded a quadrangle sheet; the game protector leaned forward to look at it over his shoulder. The sergeant ran a finger from one to another of a series of variously colored crosses which had been marked on the map.

"Monday night, over here on Copperhead Mountain, that cow was killed," he said. "The next night, about ten o'clock, that sheepflock was hit, on this side of Copperhead, right about here. Early Wednesday night, that mule got slashed up in the woods back of the Weston farm. It was only slightly injured; must have kicked the whatzit and got away, but the whatzit wasn't too badly hurt, because a few hours later, it hit that turkey–flock on the Rhymer farm. And last night, it did that." He jerked a thumb over his shoulder at the Strawmyer farm. "See, following the ridges, working toward the southeast, avoiding open ground, killing only at night. Could be a bobcat, at that."

"Or Jink's maniac with the machete," Parker agreed. "Let's go up by Hindman's gap and see if we can see anything."

* * * * *

They turned, after a while, into a rutted dirt road, which deteriorated steadily into a grass–grown track through the woods. Finally, they stopped, and the private backed off the road. The three men got out; Parker with his

Winchester, the sergeant checking the drum of a Thompson, and the private pumping a buckshot shell into the chamber of a riot gun. For half an hour, they followed the brush-grown trail beside the little stream; once, they passed a dark gray commercial-model jeep, backed to one side. Then they came to the head of the gap.

A man, wearing a tweed coat, tan field boots, and khaki breeches, was sitting on a log, smoking a pipe; he had a bolt-action rifle across his knees, and a pair of binoculars hung from his neck. He seemed about thirty years old, and any bobby-soxer's idol of the screen would have envied him the handsome regularity of his strangely immobile features. As Parker and the two State policemen approached, he rose, slinging his rifle, and greeted them.

"Sergeant Haines, isn't it?" he asked pleasantly. "Are you gentlemen out hunting the critter, too?"

"Good afternoon, Mr. Lee. I thought that was your jeep I saw, down the road a little." The sergeant turned to the others. "Mr. Richard Lee; staying at the old Kinchwalter place, the other side of Rutter's Fort. This is Mr. Parker, the district game protector. And Private Zinkowski." He glanced at the rifle. "Are you out hunting for it, too?"

"Yes, I thought I might find something, up here. What do you think it is?"

"I don't know," the sergeant admitted. "It could be a bobcat. Canada lynx. Jink, here, has a theory that it's some escapee from the paper-doll factory, with a machete. Me, I hope not, but I'm not ignoring the possibility."

The man with the matinee-idol's face nodded. "It could be a lynx. I understand they're not unknown, in this section."

"We paid bounties on two in this county, in the last year," Parker said. "Odd rifle you have, there; mind if I look at it?"

"Not at all." The man who had been introduced as Richard Lee unslung and handed it over. "The chamber's loaded," he cautioned.

"I never saw one like this," Parker said. "Foreign?"

"I think so. I don't know anything about it; it belongs to a friend of mine, who loaned it to me. I think the action's German, or Czech; the rest of it's a custom job, by some West Coast gunmaker. It's chambered for some ultra-velocity wildcat load."

The rifle passed from hand to hand; the three men examined it in turn, commenting admiringly.

"You find anything, Mr. Lee?" the sergeant asked, handing it back.

"Not a trace." The man called Lee slung the rifle and began to dump the ashes from his pipe. "I was along the top of this ridge for about a mile on either side of the gap, and down the other side as far as Hindman's Run; I didn't find any tracks, or any indication of where it had made a kill."

The game protector nodded, turning to Sergeant Haines.

"There's no use us going any farther," he said. "Ten to one, it followed that line of woods back of Strawmyer's, and crossed over to the other ridge. I think our best bet would be the hollow at the head of Lowrie's Run. What do you think?"

The sergeant agreed. The man called Richard Lee began to refill his pipe methodically.

"I think I shall stay here for a while, but I believe you're right. Lowrie's Run, or across Lowrie's Gap into Coon Valley," he said.

* * * * *

After Parker and the State policemen had gone, the man whom they had addressed as Richard Lee returned to his log and sat smoking, his rifle across his knees. From time to time, he glanced at his wrist watch and raised his head to listen. At length, faint in the distance, he heard the sound of a motor starting.

Instantly, he was on his feet. From the end of the hollow log on which he had been sitting, he produced a canvas musette–bag. Walking briskly to a patch of damp ground beside the little stream, he leaned the rifle against a tree and opened the bag. First, he took out a pair of gloves of some greenish, rubberlike substance, and put them on, drawing the long gauntlets up over his coat sleeves. Then he produced a bottle and unscrewed the cap. Being careful to avoid splashing his clothes, he went about, pouring a clear liquid upon the ground in several places. Where he poured, white vapors rose, and twigs and grass grumbled into brownish dust. After he had replaced the cap and returned the bottle to the bag, he waited for a few minutes, then took a spatula from the musette and dug where he had poured the fluid, prying loose four black, irregular–shaped lumps of matter, which he carried to the running water and washed carefully, before wrapping them and putting them in the bag, along with the gloves. Then he slung bag and rifle and started down the trail to where he had parked the jeep.

Half an hour later, after driving through the little farming village of Rutter's Fort, he pulled into the barnyard of a rundown farm and backed through the open doors of the barn. He closed the double doors behind him, and barred them from within. Then he went to the rear wall of the barn, which was much closer the front than the outside dimensions of the barn would have indicated.

He took from his pocket a black object like an automatic pencil. Hunting over the rough plank wall, he found a small hole and inserted the pointed end of the pseudo–pencil, pressing on the other end. For an instant, nothing happened. Then a ten–foot–square section of the wall receded two feet and slid noiselessly to one side. The section which had slid inward had been built of three–inch steel, masked by a thin covering of boards; the wall around it was two–foot concrete, similarly camouflaged. He stepped quickly inside.

Fumbling at the right side of the opening, he found a switch and flicked it. Instantly, the massive steel plate slid back into place with a soft, oily click. As it did, lights came on within the hidden room, disclosing a great semiglobe of some fine metallic mesh, thirty feet in diameter and fifteen in height. There was a sliding door at one side of this; the man called Richard Lee opened and entered through it, closing it behind him. Then he turned to the center of the hollow dome, where an armchair was placed in front of a small desk below a large instrument panel. The gauges and dials on the panel, and the levers and switches and buttons on the desk control board, were all lettered and numbered with characters not of the Roman alphabet or the Arabic notation, and, within instant reach of the occupant of the chair, a pistollike weapon lay on the desk. It had a conventional index–finger trigger and a hand–fit grip, but, instead of a tubular barrel, two slender parallel metal rods extended about four inches forward of the receiver, joined together at what would correspond to the muzzle by a streamlined knob of some light blue ceramic or plastic substance.

The man with the handsome immobile face deposited his rifle and musette on the floor beside the chair and sat down. First, he picked up the pistollike weapon and checked it, and then he examined the many instruments on the panel in front of him. Finally, he flicked a switch on the control board.

At once, a small humming began, from some point overhead. It wavered and shrilled and mounted in intensity, and then fell to a steady monotone. The dome about him flickered with a queer, cold iridescence, and slowly vanished. The hidden room vanished, and he was looking into the shadowy interior of a deserted barn. The barn vanished; blue sky appeared above, streaked with wisps of high cirrus cloud. The autumn landscape flickered unreally. Buildings appeared and vanished, and other buildings came and went in a twinkling. All around him, half–seen shapes moved briefly and disappeared.

Once, the figure of a man appeared, inside the circle of the dome. He had an angry, brutal face, and he wore a black tunic piped with silver, and black breeches, and polished black boots, and there was an insignia, composed of a cross and thunderbolt, on his cap. He held an automatic pistol in his hand.

Instantly, the man at the desk snatched up his own weapon and thumbed off the safety, but before he could lift and aim it, the intruder stumbled and passed outside the force–field which surrounded the chair and instruments.

For a while, there were fires raging outside, and for a while, the man at the desk was surrounded by a great hall, with a high, vaulted ceiling, through which figures flitted and vanished. For a while, there were vistas of deep forests, always set in the same background of mountains and always under the same blue cirrus–laced sky. There was an interval of flickering blue–white light, of unbearable intensity. Then the man at the desk was surrounded by the interior of vast industrial works. The moving figures around him slowed, and became more distinct. For an instant, the man in the chair grinned as he found himself looking into a big washroom, where a tall blond girl was taking a shower bath, and a pert little redhead was vigorously drying herself with a towel. The dome grew visible, coruscating with many–colored lights and then the humming died and the dome became a cold and inert mesh of fine white metal. A green light above flashed on and off slowly.

He stabbed a button and flipped a switch, then got to his feet, picking up his rifle and musette and fumbling under his shirt for a small mesh bag, from which he took an inch-wide disk of blue plastic. Unlocking a container on the instrument panel, he removed a small roll of solidograph-film, which he stowed in his bag. Then he slid open the door and emerged into his own dimension of space-time.

Outside was a wide hallway, with a pale green floor, paler green walls, and a ceiling of greenish off-white. A big hole had been cut to accommodate the dome, and across the hallway a desk had been set up, and at it sat a clerk in a pale blue tunic, who was just taking the audio-plugs of a music-box out of his ears. A couple of policemen in green uniforms, with ultrasonic paralyzers dangling by thongs from their left wrists and bolstered sigma-ray needlers like the one on the desk inside the dome, were kidding with some girls in vivid orange and scarlet and green smocks. One of these, in bright green, was a duplicate of the one he had seen rubbing herself down with a towel.

"Here comes your boss-man," one of the girls told the cops, as he approached. They both turned and saluted casually. The man who had lately been using the name of Richard Lee responded to their greeting and went to the desk. The policemen grasped their paralyzers, drew their needlers, and hurried into the dome.

Taking the disk of blue plastic from his packet, he handed it to the clerk at the desk, who dropped it into a slot in the voder in front of him. Instantly, a mechanical voice responded:

"Verkan Vall, blue-seal noble, hereditary Mavrad of Nerros. Special Chief's Assistant, Paratime Police, special assignment. Subject to no orders below those of Tortha Karf, Chief of Paratime Police. To be given all courtesies and co-operation within the Paratime Transposition Code and the Police Powers Code. Further particulars?"

The clerk pressed the "no"-button. The blue sigil fell out the release-slot and was handed back to its bearer, who was drawing up his left sleeve.

"You'll want to be sure I'm *your* Verkan Vall, I suppose?" he said, extending his arm.

"Yes, quite, sir."

The clerk touched his arm with a small instrument which swabbed it with antiseptic, drew a minute blood-sample, and medicated the needle prick, all in one almost painless operation. He put the blood-drop on a slide and inserted it at one side of a comparison microscope, nodding. It showed the same distinctive permanent colloid pattern as the sample he had ready for comparison; the colloid pattern given in infancy by injection to the man in front of him, to set him apart from all the myriad other Verkan Valls on every other probability-line of paratime.

"Right, sir," the clerk nodded.

The two policemen came out of the dome, their needlers holstered and their vigilance relaxed. They were lighting cigarettes as they emerged.

"It's all right, sir," one of them said. "You didn't bring anything in with you, this trip."

The other cop chuckled. "Remember that Fifth Level wild-man who came in on the freight conveyor at Jandar, last month?" he asked.

If he was hoping that some of the girls would want to know, what wild-man, it was a vain hope. With a blue-seal mavrad around, what chance did a couple of ordinary coppers have? The girls were already converging on Verkan Vall.

"When are you going to get that monstrosity out of our restroom," the little redhead in green coveralls was demanding. "If it wasn't for that thing, I'd be taking a shower, right now."

"You were just finishing one, about fifty paraseconds off, when I came through," Verkan Vall told her.

The girl looked at him in obviously feigned indignation.

"Why, you—You *parapeeper*!"

Verkan Vall chuckled and turned to the clerk. "I want a strato-rocket and pilot, for Dhergabar, right away. Call Dhergabar Paratime Police Field and give them my ETA; have an air–taxi meet me, and have the chief notified that I'm coming in. Extraordinary report. Keep a guard over the conveyor; I think I'm going to need it, again, soon." He turned to the little redhead. "Want to show me the way out of here, to the rocket field?" he asked.

* * * * *

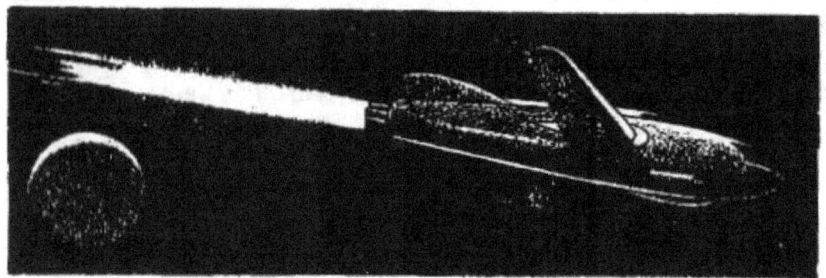

Outside, on the open landing field, Verkan Vall glanced up at the sky, then looked at his watch. It had been twenty minutes since he had backed the jeep into the barn, on that distant other time–line; the same delicate lines of white cirrus were etched across the blue above. The constancy of the weather, even across two hundred thousand parayears of perpendicular time, never failed to impress him. The long curve of the mountains was the same, and they were mottled with the same autumn colors, but where the little village of Rutter's Fort stood on that other line of probability, the white towers of an apartment–city rose—the living quarters of the plant personnel.

The rocket that was to take him to headquarters was being hoisted with a crane and lowered into the firing–stand, and he walked briskly toward it, his rifle and musette slung. A boyish–looking pilot was on the platform, opening the door of the rocket; he stood aside for Verkan Vall to enter, then followed and closed it, dogging it shut while his passenger stowed his bag and rifle and strapped himself into a seat.

"Dhergabar Commercial Terminal, sir?" the pilot asked, taking the adjoining seat at the controls.

"Paratime Police Field, back of the Paratime Administration Building."

"Right, sir. Twenty seconds to blast, when you're ready."

"Ready now." Verkan Vall relaxed, counting seconds subconsciously.

The rocket trembled, and Verkan Vall felt himself being pushed gently back against the upholstery. The seats, and the pilot's instrument panel in front of them, swung on gimbals, and the finger of the indicator swept slowly over a ninety–degree arc as the rocket rose and leveled. By then, the high cirrus clouds Verkan Vall had watched from the field were far below; they were well into the stratosphere.

There would be nothing to do, now, for the three hours in which the rocket sped northward across the pole and southward to Dhergabar; the navigation was entirely in the electronic hands of the robot controls. Verkan Vall got out his pipe and lit it; the pilot lit a cigarette.

"That's an odd pipe, sir," the pilot said. "Out–time item?"

"Yes, Fourth Probability Level; typical of the whole paratime belt I was working in." Verkan Vall handed it over for inspection. "The bowl's natural brier–root; the stem's a sort of plastic made from the sap of certain tropical trees. The little white dot is the maker's trademark; it's made of elephant tusk."

"Sounds pretty crude to me, sir." The pilot handed it back. "Nice workmanship, though. Looks like good machine production."

"Yes. The sector I was on is really quite advanced, for an electro–chemical civilization. That weapon I brought back with me—that solid–missile projector—is typical of most Fourth Level culture. Moving parts machined to the closest tolerances, and interchangeable with similar parts of all similar weapons. The missile is a small bolt of cupro–alloy coated lead, propelled by expanding gases from the ignition of some nitro–cellulose compound. Most of their scientific advance occurred within the past century, and most of that in the past forty years. Of course, the life–expectancy on that level is only about seventy years."

"Humph! I'm seventy–eight, last birthday," the boyish–looking pilot snorted. "Their medical science must be mostly witchcraft!"

"Until quite recently, it was," Verkan Vall agreed. "Same story there as in everything else—rapid advancement in the past few decades, after thousands of years of cultural inertia."

"You know, sir, I don't really understand this paratime stuff," the pilot confessed. "I know that all time is totally present, and that every moment has its own past–future line of event–sequence, and that all events in space–time occur according to maximum probability, but I just don't get this alternate probability stuff, at all. If something exists, it's because it's the maximum–probability effect of prior causes; why does anything else exist on any other time–line?"

Verkan Vall blew smoke at the air–renovator. A lecture on paratime theory would nicely fill in the three–hour interval until the landing at Dhergabar. At least, this kid was asking intelligent questions.

"Well, you know the principal of time–passage, I suppose?" he began.

"Yes, of course; Rhogom's Doctrine. The basis of most of our psychical science. We exist perpetually at all moments within our life–span; our extraphysical ego component passes from the ego existing at one moment to the ego existing at the next. During unconsciousness, the EPC is 'time–free'; it may detach, and connect at some other moment, with the ego existing at that time–point. That's how we precog. We take an autohypno and recover memories brought back from the future moment and buried in the subconscious mind."

"That's right," Verkan Vall told him. "And even without the autohypno, a lot of precognitive matter leaks out of the subconscious and into the conscious mind, usually in distorted forms, or else inspires 'instinctive' acts, the motivation for which is not brought to the level of consciousness. For instance, suppose, you're walking along North Promenade, in Dhergabar, and you come to the Martian Palace Café, and you go in for a drink, and meet some girl, and strike up an acquaintance with her. This chance acquaintance develops into a love affair, and a year later, out of jealousy, she rays you half a dozen times with a needler."

"Just about that happened to a friend of mine, not long ago," the pilot said. "Go on, sir."

"Well, in the microsecond or so before you die—or afterward, for that matter, because we know that the extraphysical component survives physical destruction—your EPC slips back a couple of years, and re‑connects at some point pastward of your first meeting with this girl, and carries with it memories of everything up to the moment of detachment, all of which are indelibly recorded in your subconscious mind. So, when you re‑experience the event of standing outside the Martian Palace with a thirst, you go on to the Starway, or Nhergal's, or some other bar. In both cases, on both time–lines, you follow the line of maximum probability; in the second case, your subconscious future memories are an added causal factor."

"And when I back–slip, after I've been needled, I generate a new time–line? Is that it?"

Verkan Vall made a small sound of impatience. "No such thing!" he exclaimed. "It's semantically inadmissible to talk about the total presence of time with one breath and about generating new time–lines with the next. *All* time–lines are totally present, in perpetual co–existence. The theory is that the EPC passes from one moment, on one time–line, to the next moment on the next line, so that the true passage of the EPC from moment to moment is a two–dimensional diagonal. So, in the case we're using, the event of your going into the Martian Palace exists on one time–line, and the event of your passing along to the Starway exists on another, but both are events in real existence.

"Now, what we do, in paratime transposition, is to build up a hypertemporal field to include the time–line we want to reach, and then shift over to it. Same point in the plenum; same point in primary time—plus primary time elapsed during mechanical and electronic lag in the relays—but a different line of secondary time."

"Then why don't we have past–future time travel on our own time–line?" the pilot wanted to know.

That was a question every paratimer has to answer, every time he talks paratime to the laity. Verkan Vall had been expecting it; he answered patiently.

"The Ghaldron–Hesthor field–generator is like every other mechanism; it can operate only in the area of primary time in which it exists. It can transpose to any other time–line, and carry with it anything inside its field, but it can't go outside its own temporal area of existence, any more than a bullet from that rifle can hit the target a week before it's fired," Verkan Vall pointed out. "Anything inside the field is supposed to be unaffected by anything outside. *Supposed to be* is the way to put it; it doesn't always work. Once in a while, something pretty nasty gets picked up in transit." He thought, briefly, of the man in the black tunic. "That's why we have armed guards at terminals."

"Suppose you pick up a blast from a nucleonic bomb," the pilot asked, "or something red–hot, or radioactive?"

"We have a monument, at Paratime Police Headquarters, in Dhergabar, bearing the names of our own personnel who didn't make it back. It's a large monument; over the past ten thousand years, it's been inscribed with quite a few names."

"You can have it; I'll stick to rockets!" the pilot replied. "Tell me another thing, though: What's all this about levels, and sectors, and belts? What's the difference?"

"Purely arbitrary terms. There are five main probability levels, derived from the five possible outcomes of the attempt to colonize this planet, seventy–five thousand years ago. We're on the First Level—complete success, and colony fully established. The Fifth Level is the probability of complete failure—no human population established on this planet, and indigenous quasi–human life evolved indigenously. On the Fourth Level, the colonists evidently met with some disaster and lost all memory of their extraterrestrial origin, as well as all extraterrestrial culture. As far as they know, they are an indigenous race; they have a long pre–history of stone–age savagery.

"Sectors are areas of paratime on any level in which the prevalent culture has a common origin and common characteristics. They are divided more or less arbitrarily into sub–sectors. Belts are areas within sub–sectors where conditions are the result of recent alternate probabilities. For instance, I've just come from the Europo–American Sector of the Fourth Level, an area of about ten thousand parayears in depth, in which the dominant civilization developed on the North–West Continent of the Major Land Mass, and spread from there to the Minor Land Mass. The line on which I was operating is also part of a sub–sector of about three thousand parayears' depth, and a belt developing from one of several probable outcomes of a war concluded about three elapsed years ago. On that time–line, the field at the Hagraban Synthetics Works, where we took off, is part of an abandoned farm; on the site of Hagraban City is a little farming village. Those things are there, right now, both in primary time and in the plenum. They are about two hundred and fifty thousand parayears perpendicular to each other, and each is of the same general order of reality."

The red light overhead flashed on. The pilot looked into his visor and put his hands to the manual controls, in case of failure of the robot controls. The rocket landed smoothly, however; there was a slight jar as it was grappled by the crane and hoisted upright, the seats turning in their gimbals. Pilot and passenger unstrapped themselves and hurried through the refrigerated outlet and away from the glowing–hot rocket.

* * * * *

An air–taxi, emblazoned with the device of the Paratime Police, was waiting. Verkan Vall said good–by to the rocket–pilot and took his seat beside the pilot of the aircab; the latter lifted his vehicle above the building level and then set it down on the landing–stage of the Paratime Police Building in a long, side–swooping glide. An express elevator took Verkan Vall down to one of the middle stages, where he showed his sigil to the guard outside the door of Tortha Karf's office and was admitted at once.

The Paratime Police chief rose from behind his semicircular desk, with its array of keyboards and viewing–screens and communicators. He was a big man, well past his two hundredth year; his hair was iron–gray and thinning in front, he had begun to grow thick at the waist, and his calm features bore the lines of middle age. He wore the dark–green uniform of the Paratime Police.

"Well, Vall," he greeted. "Everything secure?"

"Not exactly, sir." Verkan Vall came around the desk, deposited his rifle and bag on the floor, and sat down in one of the spare chairs. "I'll have to go back again."

"So?" His chief lit a cigarette and waited.

"I traced Gavran Sarn." Verkan Vall got out his pipe and began to fill it. "But that's only the beginning. I have to trace something else. Gavran Sarn exceeded his Paratime permit, and took one of his pets along. A Venusian nighthound."

Tortha Karf's expression did not alter; it merely grew more intense. He used one of the short, semantically ugly terms which serve, in place of profanity, as the emotional release of a race that has forgotten all the taboos and terminologies of supernaturalistic religion and sex–inhibition.

"You're sure of this, of course." It was less a question than a statement.

Verkan Vall bent and took cloth–wrapped objects from his bag, unwrapping them and laying them on the desk. They were casts, in hard black plastic, of the footprints of some large three–toed animal.

"What do these look like, sir?" he asked.

Tortha Karf fingered them and nodded. Then he became as visibly angry as a man of his civilization and culture–level ever permitted himself.

"What does that fool think we have a Paratime Code for?" he demanded. "It's entirely illegal to transpose any extraterrestrial animal or object to any time–line on which space–travel is unknown. I don't care if he is a green–seal thavrad; he'll face charges, when he gets back, for this!"

"He *was* a green–seal thavrad," Verkan Vall corrected. "And he won't be coming back."

"I hope you didn't have to deal summarily with him," Tortha Karf said. "With his title, and social position, and his family's political importance, that might make difficulties. Not that it wouldn't be all right with me, of course, but we never seem to be able to make either the Management or the public realize the extremities to which we are forced, at times." He sighed. "We probably never shall."

Verkan Vall smiled faintly. "Oh, no, sir; nothing like that. He was dead before I transposed to that time–line. He was killed when he wrecked a self–propelled vehicle he was using. One of those Fourth Level automobiles. I posed as a relative and tried to claim his body for the burial–ceremony observed on that cultural level, but was told that it had been completely destroyed by fire when the fuel tank of this automobile burned. I was given certain of his effects which had passed through the fire; I found his sigil concealed inside what appeared to be a cigarette case." He took a green disk from the bag and laid it on the desk. "There's no question; Gavran Sarn died in the wreck of that automobile."

"And the nighthound?"

"It was in the car with him, but it escaped. You know how fast those things are. I found that track"—he indicated one of the black casts—"in some dried mud near the scene of the wreck. As you see, the cast is slightly defective. The others were fresh this morning, when I made them."

"And what have you done so far?"

"I rented an old farm near the scene of the wreck, and installed my field–generator there. It runs through to the Hagraban Synthetics Works, about a hundred miles east of Thalna–Jarvizar. I have my this–line terminal in the girls' rest room at the durable plastics factory; handled that on a local police–power writ. Since then, I've been hunting for the nighthound. I think I can find it, but I'll need some special equipment, and a hypno–mech indoctrination. That's why I came back."

"Has it been attracting any attention?" Tortha Karf asked anxiously.

"Killing cattle in the locality; causing considerable excitement. Fortunately, it's a locality of forested mountains and valley farms, rather than a built–up industrial district. Local police and wild–game protection officers are concerned; all the farmers excited, and going armed. The theory is that it's either a wildcat of some sort, or a maniac armed with a cutlass. Either theory would conform, more or less, to the nature of its depredations. Nobody has actually seen it."

"That's good!" Tortha Karf was relieved. "Well, you'll have to go and bring it out, or kill it and obliterate the body. You know why, as well as I do."

"Certainly, sir," Verkan Vall replied. "In a primitive culture, things like this would be assigned supernatural explanations, and imbedded in the locally accepted religion. But this culture, while nominally religious, is highly rationalistic in practice. Typical lag–effect, characteristic of all expanding cultures. And this Europo–American Sector really has an expanding culture. A hundred and fifty years ago, the inhabitants of this particular time–line didn't even know how to apply steam power; now they've begun to release nuclear energy, in a few crude forms."

Tortha Karf whistled, softly. "That's quite a jump. There's a sector that'll be in for trouble, in the next few centuries."

"That is realized, locally, sir." Verkan Vall concentrated on relighting his pipe, for a moment, then continued: "I would predict space–travel on that sector within the next century. Maybe the next half–century, at least to the Moon. And the art of taxidermy is very highly developed. Now, suppose some farmer shoots that thing; what would he do with it, sir?"

Tortha Karf grunted. "Nice logic, Vall. On a most uncomfortable possibility. He'd have it mounted, and it'd be put in a museum, somewhere. And as soon as the first spaceship reaches Venus, and they find those things in a wild state, they'll have the mounted specimen identified."

"Exactly. And then, instead of beating their brains about *where* their specimen came from, they'll begin asking *when* it came from. They're quite capable of such reasoning, even now."

"A hundred years isn't a particularly long time," Tortha Karf considered. "I'll be retired, then, but you'll have my job, and it'll be your headache. You'd better get this cleaned up, now, while it can be handled. What are you going to do?"

"I'm not sure, now, sir. I want a hypno–mech indoctrination, first." Verkan Vall gestured toward the communicator on the desk. "May I?" he asked.

"Certainly." Tortha Karf slid the instrument across the desk. "Anything you want."

"Thank you, sir." Verkan Vall snapped on the code–index, found the symbol he wanted, and then punched it on the keyboard. "Special Chief's Assistant Verkan Vall," he identified himself. "Speaking from office of Tortha Karf, Chief Paratime Police. I want a complete hypno–mech on Venusian nighthounds, emphasis on wild state, special emphasis domesticated nighthounds reverted to wild state in terrestrial surroundings, extra–special emphasis hunting techniques applicable to same. The word 'nighthound' will do for trigger–symbol." He turned to Tortha Karf. "Can I take it here?"

Tortha Karf nodded, pointing to a row of booths along the far wall of the office.

"Make set–up for wired transmission; I'll take it here."

"Very well, sir; in fifteen minutes," a voice replied out of the communicator.

Verkan Vall slid the communicator back. "By the way, sir; I had a hitchhiker, on the way back. Carried him about a hundred or so parayears; picked him up about three hundred parayears after leaving my other–line terminal. Nasty–looking fellow, in a black uniform; looked like one of these private–army storm troopers you find all through that sector. Armed, and hostile. I thought I'd have to ray him, but he blundered outside the field almost at once. I have a record, if you'd care to see it."

"Yes, put it on," Tortha Karf gestured toward the solidograph–projector. "It's set for miniature reproduction here on the desk; that be all right?"

Verkan Vall nodded, getting out the film and loading it into the projector. When he pressed a button, a dome of radiance appeared on the desk top; two feet in width and a foot in height. In the middle of this appeared a small solidograph image of the interior of the conveyor, showing the desk, and the control board, and the figure of Verkan Vall seated at it. The little figure of the storm trooper appeared, pistol in hand. The little Verkan Vall snatched up his tiny needler; the storm trooper moved into one side of the dome and vanished.

Verkan Vall flipped a switch and cut out the image.

"Yes. I don't know what causes that, but it happens, now and then," Tortha Karf said. "Usually at the beginning of a transposition. I remember, when I was just a kid, about a hundred and fifty years ago—a hundred and thirty-nine, to be exact—I picked up a fellow on the Fourth Level, just about where you're operating, and dragged him a couple of hundred parayears. I went back to find him and return him to his own time–line, but before I could locate him, he'd been arrested by the local authorities as a suspicious character, and got himself shot trying to escape. I felt badly about that, but

—" Tortha Karf shrugged. "Anything else happen on the trip?"

"I ran through a belt of intermittent nucleonic bombing on the Second Level." Verkan Vall mentioned an approximate paratime location.

"Aaagh! That Khiftan civilization—by courtesy so called!" Tortha Karf pulled a wry face. "I suppose the intra–family enmities of the Hvadka Dynasty have reached critical mass again. They'll fool around till they blast themselves back to the stone age."

"Intellectually, they're about there, now. I had to operate in that sector, once—Oh, yes, another thing, sir. This rifle." Verkan Vall picked it up, emptied the magazine, and handed it to his superior. "The supplies office slipped up on this; it's not appropriate to my line of operation. It's a lovely rifle, but it's about two hundred percent in advance of existing arms design on my line. It excited the curiosity of a couple of police officers and a game–protector, who should be familiar with the weapons of their own time–line. I evaded by disclaiming ownership or intimate knowledge, and they seemed satisfied, but it worried me."

"Yes. That was made in our duplicating shops, here in Dhergabar." Tortha Karf carried it to a photographic bench, behind his desk. "I'll have it checked, while you're taking your hypno–mech. Want to exchange it for something authentic?"

"Why, no, sir. It's been identified to me, and I'd excite less suspicion with it than I would if I abandoned it and mysteriously acquired another rifle. I just wanted a check, and Supplies warned to be more careful in future."

Tortha Karf nodded approvingly. The young Mavrad of Nerros was thinking as a paratimer should.

"What's the designation of your line, again?"

Verkan Vall told him. It was a short numerical term of six places, but it expressed a number of the order of ten to the fortieth power, exact to the last digit. Tortha Karf repeated it into his stenomemograph, with explanatory comment.

"There seems to be quite a few things going wrong, in that area," he said. "Let's see, now."

He punched the designation on a keyboard; instantly, it appeared on a translucent screen in front of him. He punched another combination, and, at the top of the screen, under the number, there appeared:

EVENTS, PAST ELAPSED FIVE YEARS.

He punched again; below this line appeared the sub–heading:

EVENTS INVOLVING PARATIME TRANSPOSITION.

Another code–combination added a third line:

(ATTRACTING PUBLIC NOTICE AMONG INHABITANTS.)

He pressed the "start"–button; the headings vanished, to be replaced by page after page of print, succeeding one another on the screen as the two men read. They told strange and apparently disconnected stories—of unexplained fires and explosions; of people vanishing without trace; of unaccountable disasters to aircraft. There were many stories of an epidemic of mysterious disk–shaped objects seen in the sky, singly or in numbers. To each account was appended one or more reference–numbers. Sometimes Tortha Karf or Verkan Vall would punch one of these, and read, on an adjoining screen, the explanatory matter referred to.

Finally Tortha Karf leaned back and lit a fresh cigarette.

"Yes, indeed, Vall; very definitely we will have to take action in the matter of the runaway nighthound of the late Gavran Sarn," he said. "I'd forgotten that that was the time–line onto which the *Ardrath* expedition launched those antigrav disks. If this extraterrestrial monstrosity turns up, on the heels of that 'Flying Saucer' business, everybody above the order of intelligence of a cretin will suspect some connection."

"What really happened, in the *Ardrath* matter?" Verkan Vall inquired. "I was on the Third Level, on that Luvarian Empire operation, at the time."

"That's right; you missed that. Well, it was one of these joint–operation things. The Paratime Commission and the Space Patrol were experimenting with a new technique for throwing a spaceship into paratime. They used the cruiser *Ardrath*, Kalzarn Jann commanding. Went into space about halfway to the Moon and took up orbit, keeping on the sunlit side of the planet to avoid being observed. That was all right. But then, Captain Kalzarn ordered away a flight of antigrav disks, fully manned, to take pictures, and finally authorized a landing in the western mountain range, Northern Continent, Minor Land–Mass. That's when the trouble started."

He flipped the run–back switch, till he had recovered the page he wanted. Verkan Vall read of a Fourth Level aviator, in his little airscrew–drive craft, sighting nine high–flying saucerlike objects.

"That was how it began," Tortha Karf told him. "Before long, as other incidents of the same sort occurred, our people on that line began sending back to know what was going on. Naturally, from the different descriptions of these 'saucers', they recognized the objects as antigrav landing–disks from a spaceship. So I went to the Commission and raised atomic blazes about it, and the *Ardrath* was ordered to confine operations to the lower areas of the Fifth Level. Then our people on that time–line went to work with corrective action. Here."

He wiped the screen and then began punching combinations. Page after page appeared, bearing accounts of people who had claimed to have seen the mysterious disks, and each report was more fantastic than the last.

"The standard smother–out technique," Verkan Vall grinned. "I only heard a little talk about the 'Flying Saucers', and all of that was in joke. In that order of culture, you can always discredit one true story by setting up ten others, palpably false, parallel to it—Wasn't that the time–line the Tharmax Trading Corporation almost lost their paratime license on?"

"That's right; it was! They bought up all the cigarettes, and caused a conspicuous shortage, after Fourth Level cigarettes had been introduced on this line and had become popular. They should have spread their purchases over a number of lines, and kept them within the local supply–demand frame. And they also got into trouble with the local government for selling unrationed petrol and automobile tires. We had to send in a special–operations group, and they came closer to having to engage in out–time local politics than I care to think of." Tortha Karf quoted a line from a currently popular song about the sorrows of a policeman's life. "We're jugglers, Vall; trying to keep our traders and sociological observers and tourists and plain idiots like the late Gavran Sarn out of trouble; trying to prevent panics and disturbances and dislocations of local economy as a result of our operations; trying to keep out of out–time politics—and, at all times, at all costs and hazards, by all means, guarding the secret of paratime transposition. Sometimes I wish Ghaldron Karf and Hesthor Ghrom had strangled in their cradles!"

Verkan Vall shook his head. "No, chief," he said. "You don't mean that; not really," he said. "We've been paratiming for the past ten thousand years. When the Ghaldron–Hesthor trans–temporal field was discovered, our ancestors had pretty well exhausted the resources of this planet. We had a world population of half a billion, and it was all they could do to keep alive. After we began paratime transposition, our population climbed to ten billion, and there it stayed for the last eight thousand years. Just enough of us to enjoy our planet and the other planets of the system to the fullest; enough of everything for everybody that nobody needs fight anybody for anything. We've tapped the resources of those other worlds on other time–lines, a little here, a little there, and not enough to really hurt anybody. We've left our mark in a few places—the Dakota Badlands, and the Gobi, on the Fourth Level, for instance—but we've done no great damage to any of them."

"Except the time they blew up half the Southern Island Continent, over about five hundred parayears on the Third Level," Tortha Karf mentioned.

"Regrettable accident, to be sure," Verkan Vall conceded. "And look how much we've learned from the experiences of those other time–lines. During the Crisis, after the Fourth Interplanetary War, we might have adopted Palnar Sarn's 'Dictatorship of the Chosen' scheme, if we hadn't seen what an exactly similar scheme had done to the Jak–Hakka Civilization, on the Second Level. When Palnar Sarn was told about that, he went into paratime to see for himself, and when he returned, he renounced his proposal in horror."

Tortha Karf nodded. He wouldn't be making any mistake in turning his post over to the Mavrad of Nerros on his retirement.

"Yes, Vall; I know," he said. "But when you've been at this desk as long as I have, you'll have a sour moment or two, now and then, too."

* * * * *

A blue light flashed over one of the booths across the room. Verkan Vall got to his feet, removing his coat and hanging it on the back of his chair, and crossed the room, rolling up his left shirt sleeve. There was a relaxer-chair in the booth, with a blue plastic helmet above it. He glanced at the indicator-screen to make sure he was getting the indoctrination he called for, and then sat down in the chair and lowered the helmet over his head, inserting the ear plugs and fastening the chin strap. Then he touched his left arm with an injector which was lying on the arm of the chair, and at the same time flipped the starter switch.

Soft, slow music began to chant out of the earphones. The insidious fingers of the drug blocked off his senses, one by one. The music diminished, and the words of the hypnotic formula lulled him to sleep.

He woke, hearing the lively strains of dance music. For a while, he lay relaxed. Then he snapped off the switch, took out the ear plugs, removed the helmet and rose to his feet. Deep in his subconscious mind was the entire body of knowledge about the Venusian nighthound. He mentally pronounced the word, and at once it began flooding into his conscious mind. He knew the animal's evolutionary history, its anatomy, its characteristics, its dietary and reproductive habits, how it hunted, how it fought its enemies, how it eluded pursuit, and how best it could be tracked down and killed. He nodded. Already, a plan for dealing with Gavran Sarn's renegade pet was taking shape in his mind.

He picked a plastic cup from the dispenser, filled it from a cooler-tap with amber-colored spiced wine, and drank, tossing the cup into the disposal-bin. He placed a fresh injector on the arm of the chair, ready for the next user of the booth. Then he emerged, glancing at his Fourth Level wrist watch and mentally translating to the First Level time-scale. Three hours had passed; there had been more to learn about his quarry than he had expected.

Tortha Karf was sitting behind his desk, smoking a cigarette. It seemed as though he had not moved since Verkan Vall had left him, though the special agent knew that he had dined, attended several conferences, and done many other things.

"I checked up on your hitchhiker, Vall," the chief said. "We won't bother about him. He's a member of something called the Christian Avengers—one of those typical Europo-American race-and-religious hate groups. He belongs in a belt that is the outcome of the Hitler victory of 1940, whatever that was. Something unpleasant, I daresay. We don't owe him anything; people of that sort should be stepped on, like cockroaches. And he won't make any more trouble on the line where you dropped him than they have there already. It's in a belt of complete social and political anarchy; somebody probably shot him as soon as he emerged, because he wasn't wearing the right sort of a uniform. Nineteen–forty what, by the way?"

"Elapsed years since the birth of some religious leader," Verkan Vall explained. "And did you find out about my rifle?"

"Oh, yes. It's reproduction of something that's called a Sharp's Model '37 .235 Ultraspeed–Express. Made on an adjoining paratime belt by a company that went out of business sixty–seven years ago, elapsed time, on your line of operation. What made the difference was the Second War Between The States. I don't know what that was, either—I'm not too well up on Fourth Level history—but whatever, your line of operation didn't have it. Probably just as well for them, though they very likely had something else, as bad or worse. I put in a complaint to Supplies about it, and got you some more ammunition and reloading tools. Now, tell me what you're going to do about this nighthound business."

Tortha Karf was silent for a while, after Verkan Vall had finished.

"You're taking some awful chances, Vall," he said, at length. "The way you plan doing it, the advantages will all be with the nighthound. Those things can see as well at night as you can in daylight. I suppose you know that, though; you're the nighthound specialist, now."

"Yes. But they're accustomed to the Venus hotland marshes; it's been dry weather for the last two weeks, all over the northeastern section of the Northern Continent. I'll be able to hear it, long before it gets close to me. And I'll be wearing an electric headlamp. When I snap that on, it'll be dazzled, for a moment."

"Well, as I said, you're the nighthound specialist. There's the communicator; order anything you need." He lit a fresh cigarette from the end of the old one before crushing it out. "But be careful, Vall. It took me close to forty years to make a paratimer out of you; I don't want to have to repeat the process with somebody else before I can retire."

* * * * *

The grass was wet as Verkan Vall—who reminded himself that here he was called Richard Lee—crossed the yard from the farmhouse to the ramshackle barn, in the early autumn darkness. It had been raining that morning when the strato-rocket from Dhergabar had landed him at the Hagraban Synthetics Works, on the First Level; unaffected by the probabilities of human history, the same rain had been coming down on the old Kinchwalter farm, near Rutter's Fort, on the Fourth Level. And it had persisted all day, in a slow, deliberate drizzle.

He didn't like that. The woods would be wet, muffling his quarry's footsteps, and canceling his only advantage over the night-prowler he hunted. He had no idea, however, of postponing the hunt. If anything, the rain had made it all the more imperative that the nighthound be killed at once. At this season, a falling temperature would speedily follow. The nighthound, a creature of the hot Venus marshes, would suffer from the cold, and, taught by years of domestication to find warmth among human habitations, it would invade some isolated farmhouse, or, worse, one of the little valley villages. If it were not killed tonight, the incident he had come to prevent would certainly occur.

Going to the barn, he spread an old horse blanket on the seat of the jeep, laid his rifle on it, and then backed the jeep outside. Then he took off his coat, removing his pipe and tobacco from the pockets, and spread it on the wet grass. He unwrapped a package and took out a small plastic spray-gun he had brought with him from the First Level, aiming it at the coat and pressing the trigger until it blew itself empty. A sickening, rancid fetor tainted the air—the scent of the giant poison-roach of Venus, the one creature for which the nighthound bore an inborn, implacable hatred. It was because of this compulsive urge to attack and kill the deadly poison-roach that the first human settlers on Venus, long millennia ago, had domesticated the ugly and savage nighthound. He remembered that the Gavran family derived their title from their vast Venus hotlands estates; that Gavran Sarn, the man who had brought this thing to the Fourth Level, had been born on the inner planet. When Verkan Vall donned that coat, he would become his own living bait for the murderous fury of the creature he sought. At the moment, mastering his queasiness and putting on the coat, he objected less to that danger than to the hideous stench of the scent, to obtain which a valuable specimen had been sacrificed at the Dhergabar

Museum of Extraterrestrial Zoology, the evening before.

Carrying the wrapper and the spray–gun to an outside fireplace, he snapped his lighter to them and tossed them in. They were highly inflammable, blazing up and vanishing in a moment. He tested the electric headlamp on the front of his cap; checked his rifle; drew the heavy revolver, an authentic product of his line of operation, and flipped the cylinder out and in again. Then he got into the jeep and drove away.

For half an hour, he drove quickly along the valley roads. Now and then, he passed farmhouses, and dogs, puzzled and angered by the alien scent his coat bore, barked furiously. At length, he turned into a back road, and from this to the barely discernible trace of an old log road. The rain had stopped, and, in order to be ready to fire in any direction at any time, he had removed the top of the jeep. Now he had to crouch below the windshield to avoid overhanging branches. Once three deer—a buck and two does—stopped in front of him and stared for a moment, then bounded away with a flutter of white tails.

He was driving slowly, now; laying behind him a reeking trail of scent. There had been another stock–killing, the night before, while he had been on the First Level. The locality of this latest depredation had confirmed his estimate of the beast's probable movements, and indicated where it might be prowling, tonight. He was certain that it was somewhere near; sooner or later, it would pick up the scent.

Finally, he stopped, snapping out his lights. He had chosen this spot carefully, while studying the Geological Survey map, that afternoon; he was on the grade of an old railroad line, now abandoned and its track long removed, which had served the logging operations of fifty years ago. On one side, the mountain slanted sharply upward; on the other, it fell away sharply. If the nighthound were below him, it would have to climb that forty–five degree slope, and could not avoid dislodging loose stones, or otherwise making a noise. He would get out on that side; if the nighthound were above him, the jeep would protect him when it charged. He got to the ground, thumbing off the safety of his rifle, and an instant later he knew that he had made a mistake which could easily cost him his life; a mistake from which neither his comprehensive logic nor his hypnotically acquired knowledge of the beast's habits had saved him.

As he stepped to the ground, facing toward the front of the jeep, he heard a low, whining cry behind him, and a rush of padded feet. He whirled, snapping on the headlamp with his left hand and thrusting out his rifle pistol–wise in his right. For a split second, he saw the charging animal, its long, lizardlike head split in a toothy grin, its talon–tipped fore–paws extended.

He fired, and the bullet went wild. The next instant, the rifle was knocked from his hand. Instinctively, he flung up his left arm to shield his eyes. Claws raked his left arm and shoulder, something struck him heavily along the left side, and his cap–light went out as he dropped and rolled under the jeep, drawing in his legs and fumbling under his coat for the revolver.

In that instant, he knew what had gone wrong. His plan had been entirely too much of a success. The nighthound had winded him as he had driven up the old railroad–grade, and had followed. Its best running speed had been just good enough to keep it a hundred or so feet behind the jeep, and the motor–noise had covered the padding of its feet. In the few moments between stopping the little car and getting out, the nighthound had been able to close the distance and spring upon him.

* * * * *

It was characteristic of First–Level mentality that Verkan Vall wasted no moments on self–reproach or panic. While he was still rolling under his jeep, his mind had been busy with plans to retrieve the situation. Something touched the heel of one boot, and he froze his leg into immobility, at the same time trying to get the big Smith & Wesson free. The shoulder–holster, he found, was badly torn, though made of the heaviest skirting–leather, and the spring which retained the weapon in place had been wrenched and bent until he needed both hands to draw. The eight–inch slashing–claw of the nighthound's right intermediary limb had raked him; only the instinctive motion of throwing up his arm, and the fact that he wore the revolver in a shoulder–holster, had saved his life.

The nighthound was prowling around the jeep, whining frantically. It was badly confused. It could see quite well, even in the close darkness of the starless night; its eyes were of a nature capable of perceiving infrared radiations as light. There were plenty of these; the jeep's engine, lately running on four–wheel drive, was quite hot. Had he been standing alone, especially on this raw, chilly night, Verkan Vall's own body–heat would have lighted him up like a jack–o'–lantern. Now, however, the hot engine above him masked his own radiations. Moreover, the poison–roach scent on his coat was coming up through the floor board and mingling with the scent on the seat, yet the nighthound couldn't find the two–and–a–half foot insectlike thing that should have been producing it. Verkan Vall lay motionless, wondering how long the next move would be in coming. Then he heard a thud above him, followed by a furious tearing as the nighthound ripped the blanket and began rending at the seat cushion.

"Hope it gets a paw–full of seat–springs," Verkan Vall commented mentally. He had already found a stone about the size of his two fists, and another slightly smaller, and had put one in each of the side pockets of the coat. Now he slipped his revolver into his waist–belt and writhed out of the coat, shedding the ruined shoulder–holster at the same time. Wriggling on the flat of his back, he squirmed between the rear wheels, until he was able to sit up, behind the jeep. Then, swinging the weighted coat, he flung it forward, over the nighthound and the jeep itself, at the same time drawing his revolver.

Immediately, the nighthound, lured by the sudden movement of the principal source of the scent, jumped out of the jeep and bounded after the coat, and there was considerable noise in the brush on the lower side of the railroad grade. At once, Verkan Vall swarmed into the jeep and snapped on the lights.

His stratagem had succeeded beautifully. The stinking coat had landed on the top of a small bush, about ten feet in front of the jeep and ten feet from the ground. The nighthound, erect on its haunches, was reaching out with its front paws to drag it down, and slashing angrily at it with its single-clawed intermediary limbs. Its back was to Verkan Vall.

His sights clearly defined by the lights in front of him, the paratimer centered them on the base of the creature's spine, just above its secondary shoulders, and carefully squeezed the trigger. The big .357 Magnum bucked in his hand and belched flame and sound—if only these Fourth Level weapons weren't so confoundedly boisterous!—and the nighthound screamed and fell. Recocking the revolver, Verkan Vall waited for an instant, then nodded in satisfaction. The beast's spine had been smashed, and its hind quarters, and even its intermediary fighting limbs had been paralyzed. He aimed carefully for a second shot and fired into the base of the thing's skull. It quivered and died.

* * * * *

Getting a flashlight, he found his rifle, sticking muzzle-down in the mud a little behind and to the right of the jeep, and swore briefly in the local Fourth Level idiom, for Verkan Vall was a man who loved good weapons, be they sigma-ray needlers, neutron-disruption blasters, or the solid-missile projectors of the lower levels. By this time, he was feeling considerable pain from the claw-wounds he had received. He peeled off his shirt and tossed it over the hood of the jeep.

Tortha Karf had advised him to carry a needler, or a blaster, or a neurostat-gun, but Verkan Vall had been unwilling to take such arms onto the Fourth Level. In event of mishap to himself, it would be all too easy for such a weapon to fall into the hands of someone able to deduce from it scientific principles too far in advance of the general Fourth Level culture. But there had been one First Level item which he had permitted himself, mainly because, suitably packaged, it was not readily identifiable as such. Digging a respectable Fourth-Level leatherette case from under the seat, he opened it and took out a pint bottle with a red poison-label, and a towel. Saturating the towel with the contents of the bottle, he rubbed every inch of his torso with it, so as not to miss even the smallest break made in his skin by the septic claws of the nighthound. Whenever the lotion–soaked towel touched raw skin, a pain like the burn of a hot iron shot through him; before he was through, he was in agony. Satisfied that he had disinfected every wound, he dropped the towel and clung weakly to the side of the jeep. He grunted out a string of English oaths, and capped them with an obscene Spanish blasphemy he had picked up among the Fourth Level inhabitants of his island home of Nerros, to the south, and a thundering curse in the name of Mogga, Fire–God of Dool, in a Third–Level tongue. He mentioned Fasif, Great God of Khift, in a manner which would have got him an acid–bath if the Khiftan priests had heard him. He alluded to the baroque amatory practices of the Third–Level Illyalla people, and soothed himself, in the classical Dar–Halma tongue, with one of those rambling genealogical insults favored in the Indo–Turanian Sector of the Fourth Level.

By this time, the pain had subsided to an over–all smarting itch. He'd have to bear with that until his work was finished and he could enjoy a hot bath. He got another bottle out of the first–aid kit—a flat pint, labeled "Old Overholt," containing a locally–manufactured specific for inward and subjective wounds—and medicated himself copiously from it, corking it and slipping it into his hip pocket against future need. He gathered up the ruined shoulder–holster and threw it under the back seat. He put on his shirt. Then he went and dragged the dead nighthound onto the grade by its stumpy tail.

It was an ugly thing, weighing close to two hundred pounds, with powerfully muscled hind legs which furnished the bulk of its motive–power, and sturdy three–clawed front legs. Its secondary limbs, about a third of the way back from its front shoulders, were long and slender; normally, they were carried folded closely against the body, and each was armed with a single curving claw. The revolver–bullet had gone in at the base of the skull and emerged under the jaw; the head was relatively undamaged. Verkan Vall was glad of that; he wanted that head for the trophy–room of his home on Nerros. Grunting and straining, he got the thing into the back of the jeep, and flung his almost shredded tweed coat over it.

A last look around assured him that he had left nothing unaccountable or suspicious. The brush was broken where the nighthound had been tearing at the coat; a bear might have done that. There were splashes of the viscid stuff the thing had used for blood, but they wouldn't be there long. Terrestrial rodents liked nighthound blood, and the woods were full of mice. He climbed in under the wheel, backed, turned, and drove away.

* * * * *

Inside the paratime–transposition dome, Verkan Vall turned from the body of the nighthound, which he had just dragged in, and considered the inert form of another animal—a stump–tailed, tuft–eared, tawny Canada lynx. That particular animal had already made two paratime transpositions; captured in the vast wilderness of Fifth–Level North America, it had been taken to the First Level and placed in the Dhergabar Zoological Gardens, and then, requisitioned on the authority of Tortha Karf, it had been brought to the Fourth Level by Verkan Vall. It was almost at the end of all its travels.

Verkan Vall prodded the supine animal with the toe of his boot; it twitched slightly. Its feet were cross–bound with straps, but when he saw that the narcotic was wearing off, Verkan Vall snatched a syringe, parted the fur at the base of its neck, and gave it an injection. After a moment, he picked it up in his arms and carried it out to the jeep.

"All right, pussy cat," he said, placing it under the rear seat, "this is the one–way ride. The way you're doped up, it won't hurt a bit."

He went back and rummaged in the debris of the long–deserted barn. He picked up a hoe, and discarded it as too light. An old plowshare was too unhandy. He considered a grate–bar from a heating furnace, and then he found the poleax, lying among a pile of wormeaten boards. Its handle had been shortened, at some time, to about twelve inches, converting it into a heavy hatchet. He weighed it, and tried it on a block of wood, and then, making sure that the secret door was closed, he went out again and drove off.

An hour later, he returned. Opening the secret door, he carried the ruined shoulder holster, and the straps that had bound the bobcat's feet, and the ax, now splotched with blood and tawny cat–hairs, into the dome. Then he closed the secret room, and took a long drink from the bottle on his hip.

The job was done. He would take a hot bath, and sleep in the farmhouse till noon, and then he would return to the First Level. Maybe Tortha Karf would want him to come back here for a while. The situation on this time-line was far from satisfactory, even if the crisis threatened by Gavran Sarn's renegade pet had been averted. The presence of a chief's assistant might be desirable.

At least, he had a right to expect a short vacation. He thought of the little redhead at the Hagraban Synthetics Works. What was her name? Something Kara—Morvan Kara; that was it. She'd be coming off shift about the time he'd make First Level, tomorrow afternoon.

The claw-wounds were still smarting vexatiously. A hot bath, and a night's sleep—He took another drink, lit his pipe, picked up his rifle and started across the yard to the house.

* * * * *

Private Zinkowski cradled the telephone and got up from the desk, stretching. He left the orderly-room and walked across the hall to the recreation room, where the rest of the boys were loafing. Sergeant Haines, in a languid gin-rummy game with Corporal Conner, a sheriff's deputy, and a mechanic from the service station down the road, looked up.

"Well, Sarge, I think we can write off those stock-killings," the private said.

"Yeah?" The sergeant's interest quickened.

"Yeah. I think the whatzit's had it. I just got a buzz from the railroad cops at Logansport. It seems a track-walker found a dead bobcat on the Logan River branch, about a mile or so below MMY signal tower. Looks like it tangled with that night freight up-river, and came off second best. It was near chopped to hamburger."

"MMY signal tower; that's right below Yoder's Crossing," the sergeant considered. "The Strawmyer farm night-before-last, the Amrine farm last night—Yeah, that would be about right."

"That'll suit Steve Parker; bobcats aren't protected, so it's not his trouble. And they're not a violation of state law, so it's none of our worry," Conner said. "Your deal, isn't it, Sarge?"

"Yeah. Wait a minute." The sergeant got to his feet. "I promised Sam Kane, the AP man at Logansport, that I'd let him in on anything new." He got up and started for the phone. "Phantom Killer!" He blew an impolite noise.

"Well, it was a lot of excitement, while it lasted," the deputy sheriff said. "Just like that Flying Saucer thing."

THE END

www.ingramcontent.com/pod-product-compliance
Lightning Source LLC
Chambersburg PA
CBHW030043230526
45472CB00005B/1658